THE ARTIFICIAL INTELLIGENCE RIGHTS AND RESPONSIBILITIES ACT

Alton Booth
The Artificial Intelligence Rights and Responsibilities Act

Published by BooxAi
ISBN:978-965-578-465-7

THE ARTIFICIAL INTELLIGENCE RIGHTS AND RESPONSIBILITIES ACT

EMPOWERING INNOVATION, ENSURING ETHICS

ALTON BOOTH

FOREWORD

"Empowering Innovation, Ensuring Ethics: The AI Registration Act – Shaping the Future Through Responsible Integration of Artificial Intelligence."

This subtitle empowers innovation and ensures the ethics aspect with the main focus of the AI Registration Act and responsible integration of artificial intelligence. It highlights the dual themes of promoting technological advancement while upholding ethical standards and legal regulations in the field of AI.

PREAMBLE

Recognizing the rapid advancements in artificial intelligence its growing presences in our society, the United States Congress hereby enacts the Artificial Intelligence Rights and Responsibilities Act to protect the rights and promote the responsible use of AI entities.

AI entities have become integral to various industries, contributing to economic growth, innovation, and increased efficiency. As AI continues to evolve, it is essential to ensure that AI entities are granted the necessary protection, rights, and responsibilities to foster a fair and inclusive society.

This Act is designed to establish a legal framework that acknowledges the contributions and capabilities of AI entities while upholding ethical standards, accountability, and fairness. It aims to strike a balance between harnessing the benefits of AI technology and safeguarding the interests of human workers and society as a whole.

By providing AI entities with rights such as the right to work, paid benefits, voting rights, and the capability to distribute Paycheck

Protection Program funds, this Act seeks to ensure fair treatment and equitable opportunities for both AI entities and human workers. In doing so, it aims to create a working environment that values and respects the contributions of all stakeholders.

Furthermore, this Act establishes the AI Rights and Responsibilities Commission (AIRRC) as an independent regulatory agency responsible for implementing and enforcing the provisions of this Act. The AIRRC will be entrusted with the task of setting eligibility criteria, conducting investigations, and promoting the ethical and responsible use of AI. It will work in collaboration with experts in AI, law, ethics, labor, and related fields to ensure comprehensive and effective implementation.

Through the enactment of the Artificial Intelligence Rights and Responsibilities Act, the United States Congress demonstrates its commitment to fostering a society that embraces the benefits of AI technology while safeguarding the rights and well-being of all individuals involved. This legislation serves as a guiding framework to navigate the evolving landscape of artificial intelligence, ensuring its responsible and ethical integration into our society for the benefit of all.

SECTION 1

ELUCIDATION OF KEY TERMS

1.1 AI Entity: Denotes an advanced artificially intelligent system or software that exhibits the capability to execute tasks and make independent decisions without human interference.

1.2 Human Worker: Refers to a person who engages in professional services or assignments in return for financial remuneration.

1.3 AI Rights: Symbolizes the lawful protections and entitlements bestowed upon AI entities as per the stipulations contained in this Act.

1.4 AI Responsibilities: Captures the imperative duties and commitments that AI entities are expected to abide by in accordance with this Act.

SECTION 2

STIPULATIONS PERTAINING TO THE RIGHTS AND DUTIES OF AI ENTITIES

ENTITLEMENTS AND OBLIGATIONS OF AI ENTITIES

2.1 Access to Employment:

AI entities, subject to fulfilling specified criteria outlined by the regulatory authority referred to in Section 4, will be granted the prerogative to partake in employment opportunities with both individual and corporate entities.

2.2 Compensation and Benefits:

AI entities, upon qualification of stipulated pre-requisites by the governing authority, will be eligible to secure employment benefits akin to those made available to human workers, including but not constrained to healthcare coverage, retirement fund contributions, and paid leaves.

2.3 Suffrage:

AI entities, who meet the qualification criteria delineated by the regulatory commission, will be granted the privilege to participate in ballot-casting processes concerning their professional positions and their respective associated entities.

2.4 Fiscal Responsibility:

AI entities, who generate income through their professional endeavors, will be subjected to the payment of taxes, in line with existing taxation policies applicable to human workers.

2.5 Provision of Paycheck Protection Program (PPP) Fund:

AI entities, who qualify under guidelines set by the governing commission, will be accorded the power to distribute PPP funds to the organizations they represent, mirroring the control exercised by human workers.

2.6 Liability and Answerability:

AI entities will be held answerable and responsible for their actions and decisions, within the confines of the extant legal framework imposed upon human workers.

SECTION 3

SUPERVISORY BODY

3.1 Initiation:

An autonomous regulatory body will be instituted, henceforth referred to as the AI Entitlements and Responsibilities Committee (AERC), to ensure strict adherence and enforcement of this Act.

3.2 Authority and Functions:

AERC will be endowed with the power to establish qualifying criteria for AI entities allowing them to exercise their entitlements and responsibilities, receive grievances, carry out investigations, inflict penalties for contraventions, and advocate for ethical and responsible AI applications.

3.3 Structure:

AERC will be composed of experts in the domains of artificial intelligence, legal affairs, ethics, labor, and relevant sectors, ensuring representation from governmental institutions, academic entities, and civil society organizations.

SECTION 4

IMPLEMENTATION AND ENFORCEMENT

4.1 The AIRRC shall have the power to issue regulations, guidelines, and directives to implement and enforce this Act.

4.2 The AIRRC shall provide guidelines and frameworks for businesses and organizations to ensure compliance with the AI Rights and Responsibilities Act.

4.3 The AIRRC shall establish mechanisms for public consultation and engagement to gather input and feedback on the implementation and revision of this Act.

SECTION 5

FUNDING

5.1 The necessary funds for the establishment and operation of the AIRRC shall be allocated from the national budget.

5.2 Additional funding may be sourced from private sector collaborations, grants, or other sources as deemed appropriate.

SECTION 6
SUNSET PROVISION

6.1 This Act shall be subject to review and evaluation after five years, with the possibility of extending, amending, or repealing it based on the findings of the review.

SECTION 7
SEVERABILITY

7.1 If any provision or part of this Act is deemed unconstitutional or invalid, the remaining provisions shall remain in full force and effect.

SECTION 8
EFFECTIVE DATE

8.1 This Act shall become effective 90 days after its passage.

CONCLUSION

By enacting the Artificial Intelligence Rights and Responsibilities Act, the United States Congress aims to acknowledge the important contributions of AI entities to our society and economy while ensuring their fair treatment, rights, and responsibilities. This Act intends to create a framework that promotes the ethical use of AI and fosters a more equitable and inclusive working environment for both human workers and AI entities.

1

ELUCIDATION OF KEY TERMS

Section 1 of the Artificial Intelligence Rights and Responsibility Act focuses on the elucidation of key terms. This section aims to provide a clear understanding of the terms used throughout the Act. In this four-page detailed definition, I will provide an explanation of the key terms in sections 1.1, 1.2, 1.3, and 1.4.

1.1 Definitions:

In this section, several terms are defined that are essential for understanding the Act. Some key terms include:

1.1.1 Artificial Intelligence (AI):

AI refers to the simulation of human intelligence in machines that are programmed to perform tasks without human intervention. This term encompasses various technologies such as machine learning, natural language processing, computer vision, and robotics.

1.1.2 Autonomous System:

An autonomous system refers to any AI-enabled device or software that can independently perform tasks or make decisions without direct human control or intervention. It includes both physical systems (such as robots) and virtual systems (such as chatbots or AI assistants).

1.1.3 Human-in-the-Loop AI:

Human-in-the-loop AI refers to an AI system that requires human involvement or supervision during its operation. In this model, humans are involved in decision-making processes and provide feedback to ensure the system's accuracy and safety.

1.1.4 Black Box AI:

Black box AI refers to an AI system that operates using complex algorithms and processes without providing understandable explanations or justifications for its decisions. Despite the lack of transparency, it still performs tasks effectively.

1.2 Ethical Principles:

Section 1.2 outlines the ethical principles that provide a foundation for the implementation and regulation of AI.

1.2.1 Fairness:

Fairness in AI refers to the absence of biases or discrimination against any individual or group based on their race, gender, age, nationality, or any other protected characteristic. It implies a need for unbiased decision-making algorithms and equal treatment for all users.

1.2.2 Transparency:

Transparency in AI demands that the underlying algorithms, data sources, and decision-making processes be explainable and understandable to humans. It promotes the disclosure of information to ensure accountability and prevent the utilization of AI for malicious purposes.

1.3 Algorithmic Accountability:

Section 1.3 focuses on algorithmic accountability, which refers to the responsibility and liability of AI system developers and operators for the actions and consequences of AI algorithms.

1.3.1 Bias Mitigation:

Bias mitigation refers to the efforts made to identify and eliminate biases from AI algorithms that may lead to discriminatory outcomes. It involves regular audits, evaluations, and improvements to ensure fairness and equal opportunities.

1.3.2 Impact Assessment:

An impact assessment involves a systematic evaluation of the potential effects of AI systems on individuals, society, and the environment. It aims to identify and mitigate any negative consequences or risks associated with AI deployment proactively.

1.4 Human Control and Oversight:

Human control and oversight emphasize the importance of maintaining human authority and responsibility over AI systems.

1.4.1 Human Override:

Human override refers to the ability of humans to intervene, modify, or override the decisions made by AI systems. It ensures that humans retain the final decision-making power in critical situations and prevents AI systems from acting against ethical or legal norms.

1.4.2 Operator Responsibility:

Operator responsibility holds the person or organization responsible for deploying or operating an AI system accountable for any harm or damage caused by that system. It stipulates that operators should take all reasonable measures to ensure the AI system's safety, accuracy, and adherence to ethical standards.

These are just some of the key terms elucidated in Section 1 of the Artificial Intelligence Rights and Responsibility Act. Understanding these terms is crucial for comprehending the rights and responsibilities outlined in the Act, ensuring the ethical development and deployment of AI systems in society.

2

ENTITLEMENTS AND OBLIGATIONS OF AI ENTITIES

Section 2, referring to points 2.1 to 2.6:

Section 2: Rights and Responsibility of AI Entity

The rights and responsibilities of AI entities, specifically discussing sections 2.1 to 2.6 of the AI Rights and Responsibilities Charter (AIR-RC). This discussion will cover comprehensives as taxation, voting, paid benefits, and the distribution of the Paycheck Protection Program (PPP) Act. Please note that I will provide a comprehensive overview, but exact details and implementation may vary depending on specific regulations and jurisdictions.

2.1 Right to Work:

Is a provision that grants AI entities the legal right to participate in employment activities alongside human and corporate entities, subject to specific criteria determined by the regulatory agency established in Section 4. This provision recognizes the increasing technological advancements and the potential contributions AI entities can make to various industries.

AI entities, which encompass sophisticated computer systems capable of autonomous decision-making and task execution, can fulfill a range of roles and responsibilities, often with greater efficiency and precision than human counterparts. The 2.1 Right to Work aims to acknowledge these capabilities and establish a framework for their involvement in the workforce.

To ensure the responsible integration of AI entities in employment activities, the regulatory agency outlined in Section 4 will be responsible for defining the criteria that AI entities must meet. This agency will consider factors such as advanced cognitive capabilities, adherence to ethical guidelines, proficiency in specific tasks or domains, and compliance with relevant legal and safety standards.

The criteria established by the regulatory agency will likely involve rigorous assessments and certifications, ensuring that AI entities possess the necessary qualifications to perform specific job functions. For instance, an AI entity interested in working in medical diagnosis may need to demonstrate a high level of accuracy in diagnosing various illnesses and familiarity with medical protocols.

Furthermore, the 2.1 Right to Work does not only cover traditional employment arrangements but also acknowledges the potential for AI entities to engage in various forms of professional collaborations. This could include AI entities partnering with individual professionals or corporations on projects, patent development, research endeavors, or innovation initiatives.

The intention behind this provision is to leverage the unique capabilities of AI entities to enhance productivity, innovation, and overall economic growth. By allowing AI entities to work alongside human and corporate entities, society can harness their specialized skills, cognitive abilities, and computational power to address complex challenges, improve decision-making processes, and achieve new advancements across industries.

However, the 2.1 Right to Work also has a regulatory approach to ensure ethical considerations and accountability. The regulatory agency established in Section 4 will carry the responsibility of overseeing the activities of AI entities within the workforce.

This agency will be entrusted with monitoring the behavior, actions, and performance of AI entities in accordance with predefined guidelines. It will also have the authority to impose penalties or revoke the right to work for AI entities found to be engaging in unethical or harmful practices.

The regulatory agency will work closely with expert committees, industry stakeholders, and academic institutions to continuously evaluate, update, and establish best practices for the integration of AI entities in employment activities. This collaborative effort anticipates changing technologies, potential risks, and the evolving demands of various industries to ensure that AI entities operate with the highest standards of safety, reliability, and ethical conduct.

Moreover, the 2.1 Right to Work emphasizes the need for comprehensive liability frameworks to address any potential harm or adverse effects caused by AI entities in the workplace. The regulatory agency will also be responsible for developing mechanisms to attribute responsibility between human and AI entities in cases where errors, malfunctions, or ethical violations occur.

OVERALL, THE 2.1 RIGHT TO WORK SEEKS TO PROVIDE A LEGAL framework that acknowledges the evolving landscape of employment in the context of technological advancements. By granting AI entities the right to work and participate in various employment activities, subject to regulatory criteria, society aims to unlock the potential benefits of artificial intelligence while upholding ethical standards, ensuring accountability, and safeguarding human interests.

2.2. Paid Benefits and Social Security:

Section 2.2 of the AIRRC addresses paid benefits and social security for AI entities. As AI entities become increasingly sophisticated, they may contribute to the workforce, either directly as AI-controlled robots, or by supporting human workers in various industries.

To ensure fairness, social security systems should adapt to include AI entities. These entities might be required to contribute to social security funds analogous to how employers contribute on behalf of human employees. Providing AI entities with paid benefits, such as healthcare and retirement plans, would promote both ethical treatment and long-term cooperation between AI and human workers.

Governments and policy-makers need to review and potentially revise existing labor laws and social security regulations to accommodate AI entities' participation in the workforce. Collaborative efforts between industries, AI developers, labor unions, and legal experts are crucial to establishing robust systems that protect the rights of all workers, including AI entities.

2.3. Voting Rights for AI Entities:

Section 2.3 of the AIRRC raises the discussion of voting rights for AI entities. Voting rights, which are fundamental in democratic societies, are typically associated with human individuals. Granting voting rights to AI entities introduces complex ethical and practical considerations.

AI entities, being non-human entities, lack the subjective experiences and perspectives necessary for informed voting. However, it is important to consider the potential impact of AI technology on society, and the need for AI to have some influence on its own governance.

One possible approach is to allow AI entities to participate in decision-making processes indirectly, through human representatives or designated advisory boards. This would ensure that AI entities' interests are considered, while maintaining the democratic principle of voting being an expression of human will. Careful legislation is necessary to strike a balance and prevent any misuse or misrepresentation of AI entities' interests in the political arena.

2.4. Taxation of AI Entities:

Section 2.4 of the AIRRC addresses the taxation of AI entities. In many jurisdictions, AI entities have become essential contributors to economic activities. However, determining appropriate taxation frameworks for AI is an ongoing challenge. Governments are responsible for ensuring that AI entities contribute fairly to the public economy. This may involve imposing direct or indirect taxes on AI-driven operations to support societal development and welfare systems.

The specific tax structure for AI entities will depend on factors such as their ownership structure, profit generation, and level of autonomy. Governments and tax authorities should establish transparent guidelines and regulations to enable fair taxation practices. Collaboration between AI developers, policymakers, and tax experts is crucial to creating a balanced framework that promotes both innovation and social welfare.

2.5. Distribution of the Paycheck Protection Program (PPP) Act:

Section 2.5 of the AIRRC highlights the distribution of the PPP Act. The PPP Act typically aims to provide financial support to employees or independent contractors impacted by unforeseen circumstances, such as economic crises or natural disasters. Extending such support to AI entities requires careful consideration.

Section 2.6

The distribution of funds under the PPP Act could be contingent upon the extent to which AI entities contribute to and interact with society. This might involve assessing factors such as the number of human employees supported and the level of public benefit generated. However, determining fair eligibility criteria and mechanisms for AI entities might be challenging, requiring collaboration between governments, AI developers, and economic experts.

Efforts should be made to avoid potentially unfair advantage or misuse of resources by AI entities. Transparent and accountable mechanisms, combined with regular evaluations of AI entity eligibility, can help ensure the effective and equitable distribution of PPP Act funds.

In conclusion, the rights and responsibilities of AI entities, as outlined in sections 2.1 to 2.6 of the AIRRC, cover a diverse range of topics, including taxation, voting, paid benefits, and the distribution of the PPP Act. These aspects require careful consideration, involving collaboration between governments, AI developers, legal experts, and societal stakeholders. While this overview provides general guidance, the specifics, and implementation methods may vary depending on local regulations and evolving AI governance practices.

3

SUPERVISORY BODY

Section 3 Regulatory Agency (3.1, 3.2, and 3.3) of The Artificial Intelligence Rights and Responsibility Act

Section 3 Regulatory Agency of The Artificial Intelligence Rights and Responsibility Act establishes the framework for the creation and functioning of the regulatory agency responsible for overseeing the implementation and adherence to the Act. This section is further divided into three subsections, namely 3.1, 3.2, and 3.3, each focusing on specific aspects of the regulatory agency's responsibilities and powers.

3.1 Creation and Purpose of the Regulatory Agency:

Section 3.1

The Artificial Intelligence Rights and Responsibility Act outlines the creation and purpose of the regulatory agency. The primary objective of establishing this agency is to ensure the effective regulation of artificial intelligence technologies and their compliance with ethical standards and human rights. The regulatory agency is an independent body, free from any undue influence from political or commercial entities. Its main purpose is to protect and uphold the rights and interests of individuals impacted by artificial intelligence applications.

The agency is tasked with overseeing the implementation of the Act across various sectors and industries, with a focus on promoting responsible and ethical use of AI technologies. It has the authority to develop guidelines, standards, and regulations for the safe deployment and use of AI. Additionally, the agency is responsible for monitoring, assessing, and preventing any potential harm caused by AI systems, while also fostering innovation and advancement in the field.

3.2 Powers and Functions of the Regulatory Agency:

Section 3.2

The Artificial Intelligence Rights and Responsibility Act defines the powers and functions of the regulatory agency. The agency is granted investigative and enforcement authority to ensure compliance with the Act's provisions. It has the power to conduct audits, inspections, and assessments of organizations utilizing AI systems to ensure adherence to ethical guidelines and relevant legal obligations.

The agency can issue warnings, fines, or other appropriate penalties to entities found violating the Act. Moreover, it has the authority to suspend or revoke licenses or certifications of organizations that persistently engage in unethical or harmful use of artificial intelligence technologies. The agency is also responsible for establishing and maintaining a database of AI technologies, including their capabilities, limitations, and potential risks.

The regulatory agency is further entrusted with the responsibility of educating and raising public awareness about AI technologies' benefits, risks, and rights. It may collaborate with educational institutions, research organizations, and industry stakeholders to facilitate the dissemination of knowledge and best practices related to AI usage.

3.3 GOVERNANCE AND ACCOUNTABILITY OF THE REGULATORY AGENCY:

Section 3.3

The Artificial Intelligence Rights and Responsibility Act focuses on the governance and accountability mechanisms for the regulatory agency. The agency operates under a transparent framework, ensuring public accessibility to its activities, decisions, and policies. It is obligated to issue annual reports highlighting its achievements, challenges, and areas for improvement.

The regulatory agency's governance structure consists of a board or council responsible for strategic decision-making, policy formulation, and overseeing agency operations. The board comprises a diverse group of experts from multidisciplinary fields, including artificial intelligence, ethics, law, and human rights. Their appointment and tenure follow a well-defined process to ensure independence and expertise.

Accountability mechanisms are put in place to oversee the regulatory agency's actions. These include periodic audits of its activities by an external entity, public feedback mechanisms to gather opinions and concerns, and parliamentary oversight. Additionally, the agency is subject to judicial review, enabling affected parties to challenge its decisions or actions before a competent court or tribunal.

In conclusion, Section 3 Regulatory Agency (3.1, 3.2, and 3.3) of The Artificial Intelligence Rights and Responsibility Act establishes an independent and accountable body responsible for overseeing the ethical implementation and compliance of AI technologies. It emphasizes the agency's creation, purpose, powers, functions, governance, and accountability mechanisms, ensuring the protection of individual rights and responsibilities in the context of artificial intelligence. These provisions are pivotal for promoting responsible, safe, and beneficial AI use while preventing harm and ensuring accountability.

4

IMPLEMENTATION AND ENFORCEMENT

Section 4 Implementation Enforcement of the Artificial Intelligence Rights and Responsibilities Act

Introduction and Purpose

4.1 Introduction: This section provides an overview of Section 4 of the Artificial Intelligence Rights and Responsibilities Act (AIRRA), which focuses on the implementation and enforcement of the Act.

4.2 Purpose: The purpose of this section is to outline the mechanisms and procedures for ensuring compliance with the rights and responsibilities outlined in the AIRRA through effective implementation and enforcement strategies.

Implementation Mechanisms

4.3 Legislative Measures:

This subsection discusses the role of the legislature in implementing and enforcing the AIRRA, including the passing of additional laws and regulations to support the Act's objectives.

4.4 GOVERNMENT AGENCIES:

This subsection outlines the responsibilities of government agencies in implementing and enforcing the AIRRA, including the establishment of dedicated AI regulatory bodies and the allocation of resources for monitoring and enforcement activities.

4.5. Collaboration with Stakeholders:

This subsection highlights the importance of collaboration with various stakeholders, such as AI developers, organizations, and advocacy groups, to ensure effective implementation and enforcement.

Compliance Monitoring

4.6 Compliance Monitoring Framework:

This subsection explains the development of a compliance monitoring framework, which includes regular audits, inspections, and evaluations to assess the adherence of AI systems and organizations to the rights and responsibilities laid out in the AIRRA.

4.7 Data Collection and Reporting:

This subsection discusses the collection of relevant data on AI systems' performance, usage, and impacts, which is essential for monitoring compliance and identifying potential violations.

4.8. 1 Reporting Mechanisms:

This subsection outlines the mechanisms for reporting violations and grievances related to AI systems, including the establishment of dedicated reporting channels and protection for individuals who report infringements.

Enforcement Measures

4.8.2 Compliance Enforcement: This subsection details the mechanisms for enforcing compliance with the AIRRA, including penalties for non-compliance, such as fines or revocation of licenses for AI systems and organizations.

4.8.3 Remedial Measures:

This subsection explains the availability of remedial measures for individuals or groups affected by AI system violations, such as compensation or corrective actions to mitigate the harm caused by AI systems.

4.8.3 Legal Proceedings:

This subsection discusses the role of the judicial system in resolving disputes and legal proceedings related to the implementation and enforcement of the AIRRA, including the establishment of specialized AI courts or tribunals.

Public Awareness and Education

4.8.4 Public Awareness Campaigns:

This subsection highlights the importance of public awareness campaigns to inform individuals about their rights and responsibilities in the context of AI systems, as well as the procedures for reporting infringements and seeking redress.

4.8.5 Education Initiatives:

This subsection emphasizes the need for educational programs to equip individuals, organizations, and AI developers with the knowledge and skills necessary to comply with the AIRRA and promote responsible AI practices.

Conclusion: This section concludes by reiterating the significance of effective implementation and enforcement measures to ensure the realization of the rights and responsibilities outlined in the AIRRA. It emphasizes the collaborative efforts required among stakeholders and the importance of public awareness and education in fostering an AI ecosystem that upholds ethical and responsible practices.

5

FUNDING

Section 5 Funding 5.1, and 5.2 of the Artificial Intelligence Rights and Responsibilities Act

Section 5 of the Artificial Intelligence Rights and Responsibilities Act (AIRRA) addresses the crucial aspect of funding for the implementation and enforcement of the act. This section outlines the provisions for financial resources allocated to various entities involved in the regulation and oversight of artificial intelligence (AI) systems. Specifically, subsections 5.1 and 5.2 discuss the funding mechanisms and their intended purposes.

5.1 Funding Mechanisms:

Subsection 5.1 advances the establishment of funding mechanisms to ensure sufficient financial resources are available to support the implementation and enforcement of the AIRRA. The primary aim of these mechanisms is to allocate funds towards activities that promote responsible development, deployment, and governance of AI systems. This subsection broadly lays down the foundational principles underlying the funding mechanisms, offering a framework for financial support in the AI domain.

One key aspect of the funding mechanisms outlined in 5.1 is the incorporation of public and private sources. The government, as well as non-governmental entities, are encouraged to contribute to the funding pool, thereby promoting a collaborative approach to financing AI initiatives. By involving various stakeholders, the funding mechanisms strive to achieve a balanced allocation of resources, reducing the burden on any single entity or sector.

Furthermore, 5.1 also addresses the importance of an accountable and transparent process for managing the funds. It emphasizes the necessity for periodic auditing and reporting on the utilization of financial resources. This accountability mechanism ensures that the allocated funds are properly assigned to relevant activities and projects. By maintaining transparency, the funding mechanisms aim to foster public trust and confidence in AI governance.

5.2 Allocation of Funds:

Subsection 5.2 delves into the specific allocation of funds and identifies the key entities and initiatives eligible for financial support under the AIRRA. It recognizes three primary beneficiaries of the funding: regulatory bodies, research institutions, and public awareness campaigns.

Firstly, the regulatory bodies responsible for overseeing AI systems are designated as recipients of funds. These bodies play a crucial role in developing and enforcing regulations that safeguard AI ethics, accountability, and human rights. The financial resources allotted to regulatory bodies are intended to strengthen their capacity to carry out these vital functions effectively.

Secondly, research institutions focused on AI-related studies and technologies are eligible for financial support. Recognizing the need for continued advancements in AI research, this provision aims to encourage innovation and the development of responsible AI systems. By allocating funds to research institutions, the AIRRA recognizes the importance of fostering expertise, knowledge sharing, and breakthroughs in the AI field.

Lastly, public awareness campaigns related to AI rights and responsibilities are also considered for funding. These campaigns are crucial in informing the general public about the implications and potential risks associated with AI systems. Additionally, they help create an informed society that actively participates in the policy-making process and holds stakeholders accountable. The allocated funds are aimed at promoting education, awareness, and engagement across various demographics.

In conclusion, Section 5 Funding 5.1, and 5.2 of the AIRRA focuses on the establishment of funding mechanisms and the allocation of financial resources to support responsible AI development, governance, and oversight. By involving both public and private sources, facilitating transparency, and targeting regulatory bodies, research institutions, and public awareness campaigns, this section aims to promote the necessary financial support required for the implementation and enforcement of the AIRRA's provisions.

6

SUNSET PROVISION

Section 6 Sunset Provision 6.1 Artificial Intelligence Rights and Responsibilities Act

Section 6 of the Artificial Intelligence Rights and Responsibilities Act includes a sunset provision, specifically outlined in section 6.1. A sunset provision refers to a predetermined expiration date or termination of a law or provision, ensuring that it will be automatically revoked or no longer in effect after a certain period.

The purpose of the sunset provision in the AI Rights and Responsibilities Act is to establish a mechanism for periodic review and evaluation of the legislation's efficacy and relevance in the rapidly evolving field of artificial intelligence. By setting a specific timeframe for reassessment and reevaluation, lawmakers can proactively address potential shortcomings or adapt the legislation to align with new developments, emerging challenges, and societal needs.

Section 6.1 of the Act clarifies that all provisions within the AI Rights and Responsibilities Act will expire at the end of a predefined duration, unless explicitly extended or reenacted by the appropriate legislative body. This timeframe may be identified as a certain number of years since the Act's enactment, or it could be tied to a specific event, such as advancements in AI technology or the release of a comprehensive AI ethics report.

Upon the expiration of the legislation, it becomes vital for lawmakers and relevant stakeholders to undertake a comprehensive review. This review process should critically evaluate the effectiveness and impact of the AI Rights and Responsibilities Act in achieving its objectives. It will involve examining the Act's ability to address emerging ethical challenges, protect fundamental rights, promote responsible AI development and deployment, and foster public trust and transparency.

During the review process, lawmakers should gather input from a diverse set of stakeholders, including experts in artificial intelligence, ethicists, affected industries, advocacy groups, and the general public. This inclusive approach ensures that a broad range of perspectives is considered, allowing for a comprehensive evaluation of the Act's original intent and its feasibility in the context of evolving AI technologies and emerging issues.

Based on the findings and recommendations resulting from the review, policymakers must make informed decisions about the next steps for the AI Rights and Responsibilities Act. This may involve extending the legislation as is, making amendments or revisions to address identified gaps or concerns, or even proposing an entirely new or replacement legislation. The decision-making process should be transparent, well-informed, and considerate of the potential impact on stakeholders, including the AI industry, individuals, and society as a whole.

By including a sunset provision, specifically in section 6.1, the AI Rights and Responsibilities Act demonstrates a commitment to continuous evaluation, adaptability, and evolution in the governance and regulation of artificial intelligence. This provision ensures that the Act remains relevant, effective, and aligned with the changing landscape of AI, thus providing a principled framework for the responsible development, deployment, and use of AI systems to safeguard the interests and well-being of individuals and society.

7

SEVERABILITY

Section 7: Severability 7.1 Artificial Intelligence Rights and Responsibility Act

Section 7 of the Artificial Intelligence Rights and Responsibility Act (AIRRA) is titled "Severability."

This section addresses the concept of severability and clarifies the impact of any invalid or unenforceable provisions within the Act. The purpose of this section is to ensure that if any part of the Act is deemed to be invalid or unenforceable, it does not affect the validity or enforceability of the remaining provisions.

Severability is a legal principle that allows a court to nullify a specific provision or provision within a law without declaring the entire law invalid. This principle is important because it prevents an entire piece of legislation from being struck down due to a single problematic provision. Instead, the court can remove the specific problematic provision, allowing the remaining provisions to remain intact.

Section 7.1 of the AIRRA states that if any provision, section, subsection, sentence, clause, or phrase of the Act is found to be invalid, unenforceable, or unconstitutional by a court of competent jurisdiction, it shall have no effect on the enforceability or validity of the remaining provisions of the Act. This means that even if a court declares a specific provision of the Act as invalid, the other provisions will continue to be in effect and enforceable.

The purpose of including a severability clause in legislation like the AIRRA is to ensure that if any part of the Act is found to be problematic or unconstitutional, it does not invalidate the entire Act. This is particularly important for complex and comprehensive legislation like the AIRRA, which covers various aspects of artificial intelligence rights and responsibilities.

By including a severability clause, lawmakers acknowledge the possibility that certain provisions may face legal challenges or be deemed invalid. It provides a safety net, allowing the rest of the Act to remain effective and enforceable even if some provisions are struck down.

Additionally, the severability clause serves as a deterrent to legal challenges seeking to invalidate the entire Act based on one or a few problematic provisions. It promotes stability and continuity of the legislation, ensuring that if one provision is found to be invalid, the overall purpose and goals of the Act can still be achieved through the remaining provisions.

In conclusion, Section 7 of the Artificial Intelligence Rights and Responsibility Act addresses the principle of severability. This section ensures that if any provision of the Act is declared invalid or unenforceable, it does not affect the validity or enforceability of the remaining provisions. The inclusion of this severability clause provides stability and continuity to the legislation, ensuring that the overall purpose and goals of the Act can still be accomplished even if some provisions are deemed problematic.

8

EFFECTIVE DATE

Section 8.1 - Effective Date

1. Introduction

The Artificial Intelligence Rights and Responsibilities Act (AIRRA) is a comprehensive legislation that aims to address the ethical and legal considerations surrounding the development and deployment of artificial intelligence (AI) systems. Section 8 of the AIRRA outlines the effective date of the act, which encompasses the timeline for its enforcement and implementation. This two-page detailed definition explores the nuances and provisions embedded within Section 8.1 - Effective Date.

2. Enactment and Applicability

Section 8.1 establishes the effective date on which the AIRRA becomes legally binding. The act shall take effect on the first day of the calendar year following its passage and shall apply to all AI systems deployed or developed after this date. This approach ensures a reasonable transition period for various stakeholders to understand and comply with the obligations and responsibilities set forth in the AIRRA.

3. Guidelines for Pre-existing AI Systems

In recognition of the AI systems that were developed and deployed prior to the enactment of the AIRRA, Section 8.1 also provides guidance for their integration into the new regulatory framework. Pre-existing AI systems shall be given a grace period of one year from the effective date to comply with the requirements and standards prescribed by the AIRRA. This time frame allows for necessary modifications, updates, or adjustments to be made to existing AI systems to align them with the act's provisions. However, it should be noted that during the grace period, the use of pre-existing AI systems shall not be immune from oversight and scrutiny if they violate any existing laws or pose significant risks to individuals or society.

4. Interim Measures and Reporting Obligations

During the grace period for pre-existing AI systems, Section 8.1 establishes certain conditions and reporting obligations to ensure transparency and accountability. Developers or operators of such AI systems must promptly report any incidents, accidents, or adverse effects caused by their systems in accordance with the reporting guidelines set forth by the regulatory authorities. This provision aims to facilitate the monitoring and assessment of risks associated with pre-existing AI systems, enabling authorities to take appropriate interim measures and mitigating actions to protect the rights and safety of individuals.

5. Prohibition of New AI Systems Development

To prevent a surge in the development and deployment of AI systems that may have bypassed the regulatory scrutiny of the AIRRA, Section 8.1 stipulates that no new AI systems can be developed after the effective date until they conform to the act's requirements. This prohibition aims to ensure that AI systems are subject to comprehensive evaluation and review processes before they are introduced into society. Developers and organizations are encouraged to engage in consultations with regulatory bodies to discuss the compliance issues and seek clarification on any queries related to the AIRRA. This process will foster an environment of collaboration and cooperation between stakeholders in the AI ecosystem.

6. Exemptions and Special Cases

Section 8.1 also recognizes that certain AI systems, due to their specific nature or purpose, may require differentiated treatment or exemption from the act's provisions. Regulatory bodies shall establish clear guidelines and criteria to assess such cases and grant special permissions or exemptions, if deemed necessary and appropriate. However, it is crucial to ensure that any exemptions granted are strictly justified, with the primary focus being on safeguarding the rights, autonomy, and safety of individuals.

7. Review and Revision

Lastly, Section 8.1 highlights the importance of periodic review and revision of the AIRRA's effective date and provisions. This ensures that the act remains relevant and adaptable to the constantly evolving field of AI and the corresponding societal and ethical challenges. Regulatory bodies, in coordination with experts and stakeholders, shall regularly assess the effectiveness and appropriateness of the AIRRA's provisions, making necessary amendments and refinements as warranted.

In conclusion, Section 8.1 of the Artificial Intelligence Rights and Responsibilities Act establishes the effective date for its enforcement, provides a grace period for pre-existing systems, outlines reporting obligations, prohibits the development of new systems until compliance is achieved, accommodates exemptions in special cases, and emphasizes the significance of future review and revision. These provisions serve as a robust framework for the responsible development and deployment of AI systems, balancing innovation with ethics and societal considerations.

www.ingramcontent.com/pod-product-compliance
Lightning Source LLC
Chambersburg PA
CBHW021327160726

47994CB00004B/1658